LEVEL 1 · SCIENCE

LET'S READ AND FIND OUT

THE ARCTIC FOX'S JOURNEY

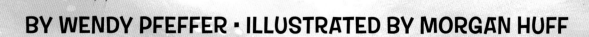

BY WENDY PFEFFER · ILLUSTRATED BY MORGAN HUFF

HARPER

An Imprint of HarperCollinsPublishers

Special thanks to Dr. Knut Kielland, Professor of Biology and Wildlife
at the Institute of Arctic Biology, for his valuable assistance.

My sincere thanks to Tamar Mays for her expert and much needed guidance.

The Let's-Read-and-Find-Out Science book series was originated by Dr. Franklyn M. Branley, Astronomer
Emeritus and former Chairman of the American Museum of Natural History–Hayden Planetarium, and was formerly
co-edited by him and Dr. Roma Gans, Professor Emeritus of Childhood Education, Teachers College, Columbia
University. Text and illustrations for each of the books in the series are checked for accuracy by an expert in the
relevant field. For more information about Let's-Read-and-Find-Out Science books, write to HarperCollins Children's
Books, 195 Broadway, New York, NY 10007, or visit our website at www.letsreadandfindout.com.

Library of Congress Control Number: 2018964887
ISBN 978-0-06-249083-4 (trade bdg.) — ISBN 978-0-06-249082-7 (pbk.)

The artist used Adobe Photoshop to create the digital illustrations for this book.
Typography by Honee Jang
19 20 21 22 23   SCP   10 9 8 7 6 5 4 3 2 1
❖
First Edition

For Caden Leeds Kianka. May he grow up
enjoying nature as much as his dad.—W.P

For Mom and Mi-mi—M.H.

During spring and summer arctic foxes live in family groups, but in the fall each fox goes off on its own. During winter's long, cold, dark days some arctic foxes take dangerous but amazing journeys.

One fox starts her winter wanderings in late fall. She roams alone across the **tundra**, a rocky, windswept, treeless plain.

As colder weather sets in, the fox's warm coat slowly turns from brown to winter white.

That's good. It **camouflages** her from hungry arctic wolves and polar bears.

The arctic fox is on the prowl. Suddenly her ears twitch. She hears the hustle and bustle of her favorite food, a fat, furry lemming. These little rodents scurry around under the snow.

The fox leaps in the air, then plunges through the lemming's snowy roof. She dives under the snow to grab dinner.

But the little lemming darts through a network of tunnels, out of the fox's reach.

The arctic fox plods away hungry. Snow swirls. The wind howls. The air is freezing cold.

Finding food is not easy. Lemmings are scarce and summer berries are gone.

Still hungry, the fox does something brave to find food. Although she's only as big as a house cat, she follows a huge polar bear. The bear spots its favorite food, a young ringed seal, swimming under the water.

The seal heads for a hole in the ice to come up for air. The bear heads to the ice and drags the seal out of the hole.

The bear eats only the seal's fat.
The fox waits patiently.

After eating its fill,
the bear lumbers off.

The hungry fox hurries over and munches
on morsels of seal meat. She eats until she is
full and only a few bones are left.

The fox heads farther north into the wilderness, toward the top of the world.

The arctic fox does not **migrate** south to warmer places.

She does not **hibernate** for the winter, resting in a safe place. Instead, she faces starvation, bitter cold, blinding **blizzards**, and gusty winds sweeping across the sea ice.

No one knows why she heads north. In this harsh environment, her journey is full of dangers as she treks over ice-choked waters, across frigid **glaciers**, and around rocky, snow-covered peaks.

On one steep cliff, the hungry fox spies an egg on the edge of a ledge. The egg was laid last summer and now is frozen in the rock-hard ice.

Clawing at the egg, the fox loosens tiny bits to eat.

Weeks later, still searching for food, she senses something ahead.
In the dark she sees a huge object. It's not moving. She hurries and finds
a dead musk ox. Animals, probably arctic wolves, had devoured most of
its meat. The fox scrapes the bones and feeds on a few frozen scraps.

21

For the next three months the sun never rises. Winter is one long night. Sometimes the **northern lights** give off a green, red, or blue glow. They swirl around and light up the sky.

The fox growls and howls, filling the night air with an eerie song. Surrounded by an icy world, the fox looks like a little ghost, with her snow-white fur blowing in the wind.

The wind picks up and violently tosses sharp pieces of ice as big as **hailstones** at the fox.

DID YOU KNOW?

Arctic winters get bitterly cold, often 30 degrees below zero Fahrenheit (-34.4 degrees Celsius). People can freeze. But not the arctic fox. She can even survive at 45 degrees below zero Fahrenheit (-43°C)— the coldest Arctic temperature—because she has a winter coat of incredibly warm, thick fur. Even her small ears and feet are covered with thick fur.

Behind a snowy mound on the ice-covered ground, she curls into a ball, sheltered from the powerful wind. Her long, bushy tail covers her nose and acts as a scarf.

Patiently, she waits out the blizzard.

When the storm passes, ice crackles. **Icebergs** moan and groan. The moon hangs low in the sky.

The fox starts to roam again, cutting a trail in deep snow. She looks like a snowball, blowing sideways and rolling around in the wind.

During the long, dark winter, the fox wanders over ice and scrambles onto a huge glacier. She jabs her claws into the ice to keep from slipping.

Finally, she treks off the glacier and heads toward a craggy, snow-covered peak, where she wanders over the ice.

The fox has wandered in a big circle and has traveled almost 1,700 miles.

As the days pass, changes begin happening. The fox can tell that she should keep heading south toward the tundra where she began her journey. Each day the sun shines a little longer.

As the sun warms the air, icebergs rumble. Some break and tumble into the sea. The fox sees moss growing on the rocks. Birds returning to the tundra peck at the moss.

One day, the fox slides down an icy hill and falls into fluffy snow. Hearing hustle and bustle beneath her, the fox dives down, and this time, she catches a fat lemming.

What joy! Spring has returned to the tundra, and so has the arctic fox.

In six months she traveled over 2,000 miles and found her way back to the tundra without any landmarks to guide her.

The arctic fox accomplished something amazing.

She survived.

AUTHOR'S NOTE

This journey is typical of ones that some arctic foxes have taken in winter. Knowledge of these comes from a Norwegian explorer's notes written over a hundred years ago, as well as researchers tagging foxes about fifty years ago and Canadian biologists recently using satellite monitors. Global warming may soon put a stop to these journeys as sea ice gradually disappears and the foxes have no way to travel from one landmass to another.

Camouflage makes an object hard to see. An arctic fox's brown fur in summer is hard to see on the brown ground of the tundra. The fox's white fur in winter is hard to see on white snow and ice. Camouflage makes it hard for hungry wolves and polar bears to see the fox.

What you need:

Brown, gray, and white paper

What you do:

1. Take brown paper, gray paper, and white paper outdoors.

2. Find a large gray stone or a stone wall. Put each paper against it. Stand back.

   - Which color paper is hardest to see against the gray stone? Why?

   - Are the white and brown papers easy or hard to see on the gray stone? Why?

3. Find a white sidewalk or something else white. Put each paper against it.

   - Stand back. Which color paper is hardest to see against the white object? Why?

   - Are the gray and brown papers hard to see or easy to see on the white object? Why?

4. Find a tree with a brown trunk. Put each paper up against the trunk. Stand back.

   - Which color paper is hardest to see against the tree trunk? Why?

   - Is this like a brown fox on the brown tundra being hard to see?

   - Is the brown fox camouflaged?

What else can you do?

1. Go inside and find things brown, gray, and white.

2. Put the three colored papers on top of each one.

3. What did you learn?

   - Did putting brown paper up against something brown blend in with it?

   - Did putting gray paper on something gray make it harder to see?

   - Did holding white paper on something white camouflage it?

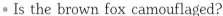

GLOSSARY

Blizzard—A violent winter storm with cold temperatures, strong winds, and lots of blowing snow.

Camouflage—To hide something or make it harder to see by changing the way it looks. The arctic fox's white coat camouflages her from hungry arctic wolves.

Glacier—Huge masses of thick ice that remain frozen from year to year. Glaciers flow like very slow rivers and can be hundreds of thousands of years old.

Hailstone—Heavy ice that falls from the clouds.

Hibernate—Resting in a safe place in the winter.

Iceberg—A large mass of ice that floats in the ocean after breaking off from a glacier.

Migrate—To move from one habitat to another. Some animals migrate when seasons change.

Northern lights—Colorful lights sometimes seen in the arctic sky. They are also called the aurora borealis.

Tundra—A large area of flat land in northern parts of the world where there are no trees and the ground is always frozen.

PARTS OF THE FOX'S ANATOMY THAT HELP HER SURVIVE THE JOURNEY

The arctic fox's body was made for snow and cold.

Fat gives emergency energy and insulation against heat loss.

Arctic fox fur is remarkably warm, including a thick, dense undercoat that keeps the cold out.

A compact body has less surface area, so there is less heat loss.

Short limbs have less surface area, so there is less heat loss.

Spike-like hairs on the bottom of an arctic fox's paws grip ice and keep them from sinking in the snow.

The fox's tail is thick and bushy and can be used like a warm scarf or cozy bed.

Be sure to look for all of these books in the **Let's-Read-and-Find-Out** Science series: